捣蛋猫爱数学

中英双语

什么是时间

A Second, a Minute, a Week with Days in It

〔美〕布赖恩·P.克利里◎著　〔美〕布赖恩·盖布尔◎绘　许晓晴◎译

时间：
一个特定的时刻，或有起点和终点的一段时间。

北京科学技术出版社

时间可以用**秒**、**分钟**、

天、**星期**、

月或者**年**来计量。

手表、日历、

手机、电脑

以及嘀嗒嘀嗒响的机械钟表，

都可以告诉你现在的时间。

1秒很短,
差不多是拍两下手、
打一个嗝或者打一个喷嚏所花的**时间**。

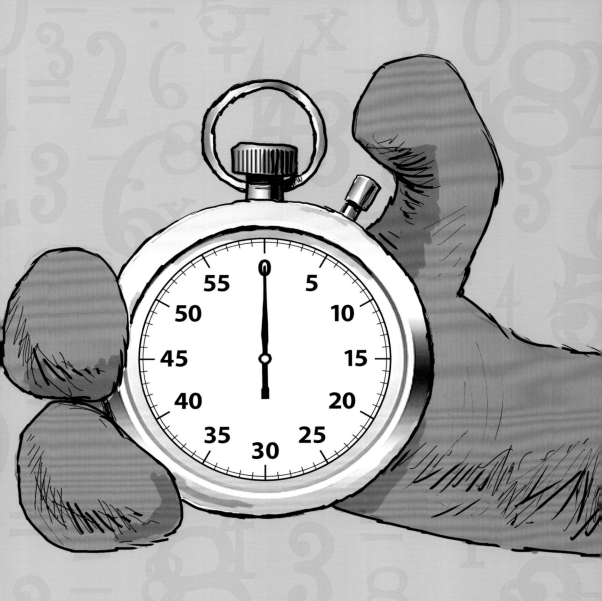

1秒也是你说出
"一、一千"所花的**时间**。
1分钟有60秒。

分钟也是**时间**单位，1分钟等于60**秒**。

跑完所有的垒或者爬40级台阶要多长时间呢?

这些大约都需要1**分钟**。

在国旗下连续宣誓4次，

或者唱4遍《祝你生日快乐》

60 **分钟**

等于1**小时**。

你的午休时间大概就有这么长。

如果你骑了1**小时**自行车，

或者滑了1**小时**冰，

你的双腿肯定很有力。

1天有24小时。

午夜12点过后,新的一天就开始了。

所以不管今天是星期一、
星期三还是星期天，
24**小时**之后，这一天就结束了。

连续的7**天**,

我们管它叫1个**星期**。

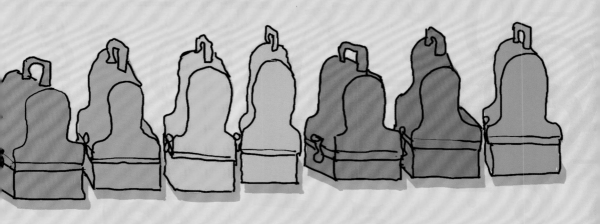

所以,一天接着一天,

1个星期的时间里你会吃7顿午餐,说7次晚安。

1个**月**有30**天**或31**天**,
还有一个月——二月——

只有28**天**或者29**天**（每4**年**一次），

我们从日历上能看到。

连续的12个**月**就是1**年**。

在1**年**的时间，

也就是52个**星期**里，

你会遇到新的老师，会有更多的家庭作业。

你生日蛋糕上的蜡烛会比去年的多1根。

10**年**为1个**年代**。

1个**年代**过去后,一个5岁的孩子就15岁了。

现在,让我们总结一下学到的知识吧!

1个**年代**有10年,
1年有52个**星期**,
1个**星期**
有连续的7**天**。

你每一天的24小时

都是在学习、玩耍和睡觉中度过的。

每一个**小时**都包含60**分钟**,

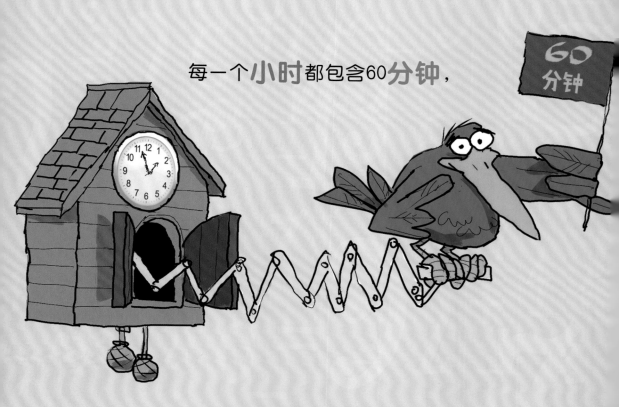

每一**分钟**都包含60**秒**。

我本来想多讲一些，
但是时间似乎已经被用完了。

A Second, a Minute, a Week with Days in It

第2~3页

Time can be measured in seconds, in minutes, in days, or in weeks, months, or years by watches or calendars, cell phones, computers, or clocks that ticktock with their gears.

第4~5页

1 second is short, like the time that it takes to clap twice or hiccup or sneeze. It's the time that you need to recite "one, one thousand." A minute has 60 of these.

第6~7页

A minute is also a unit of time, and 60 short seconds are in it. To run all the bases or climb 40 stairs? These each might be done in a minute.

第8~9页

The Pledge of Allegiance 4 times in a row or the birthday song 4 times repeated—both would take close to a minute before the speaking or song was completed.

第10~11页

It takes 60 minutes to make up 1 hour. Your lunch break might just be that long. If you rode your bike or you skated an hour, your legs would sure have to be strong!

第12~13页

See the two hands on the face of that clock in the square, way up high on the tower? When the long one goes all the way round in a circle, the time that has passed is 1 hour.

第14~15页

A day is a unit of 24 hours. At midnight, a new day's begun. So whether it's Monday or Wednesday or Sunday, in 24 hours, it's done.

第16~17页

7 straight days, flowing one to the next is a measure of time called 1 week. So that's 7 lunches and 7 good-nights, each day after day—in a streak.

> 第18~19页

A month might be 30 or 31 days, and one—February—appears on calendars either with 28 days or 29 (every 4 years).

> 第20~21页

12 months in a row is called 1 year of time. And during that year, you will handle new teachers, more homework—that's 52 weeks!—and your cake will add one birthday candle.

> 第22~23页

10 years as a group is a decade of time. A 5-year-old child will have turned 15 years old when a decade's complete! Now let's sum up the facts that we've learned.

> 第24~25页

A decade's 10 years, each with 52 weeks, each week having 7 straight days. These days are each made up of 24 hours of sleeping and schoolwork and play.

> 第26~27页

All of these hours contain 60 minutes, and inside each minute, you'll find that there're 60 seconds, and I'd explain more, but it seems that I've run out of time!

> 第28~29页

So how do we measure time?
Do you know?

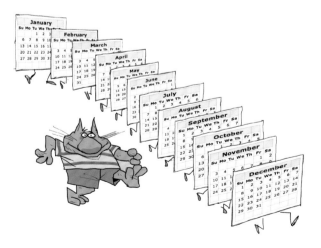

31

献给佩内洛普·米勒-史密斯和海内尔贝格小学的孩子们。
——布赖恩·P.克利里

Text copyright © 2013 by Brian P. Cleary
Illustrations copyright © 2013 by Lerner Publishing Group, Inc.
Simplified Chinese rights arranged through CA-LINK International LLC (www.ca-link.com)
English and Chinese bilingual translation © 2019 by Beijing Science and Technology Publishing Co., Ltd.

著作权合同登记号　图字：01-2018-2342

图书在版编目(CIP)数据

捣蛋猫爱数学.什么是时间/(美)布赖恩·P.克利里著；(美)布赖恩·盖布尔绘；许晓晴译.—北京：北京科学技术出版社，2019.3
　ISBN 978-7-5714-0030-9

Ⅰ.①捣… Ⅱ.①布…②布…③许… Ⅲ.①数学-儿童读物 Ⅳ.①O1-49

中国版本图书馆CIP数据核字（2019）第003210号

捣蛋猫爱数学.什么是时间

作　　者：	〔美〕布赖恩·P.克利里	绘　者：	〔美〕布赖恩·盖布尔
译　　者：	许晓晴	策划编辑：	石　婧
责任编辑：	樊川燕	责任印制：	张　良
出 版 人：	曾庆宇	出版发行：	北京科学技术出版社
社　　址：	北京西直门南大街16号	邮政编码：	100035
电话传真：	0086-10-66135495（总编室）		0086-10-66113227（发行部）
	0086-10-66161952（发行部传真）		
电子信箱：	bjkj@bjkjpress.com	网　　址：	www.bkydw.cn
经　　销：	新华书店	印　　刷：	北京宝隆世纪印刷有限公司
开　　本：	710mm×1000mm 1/16	印　　张：	2
版　　次：	2019年3月第1版	印　　次：	2019年3月第1次印刷
ISBN 978-7-5714-0030-9 /O·031			

定价：25.00元

京科版图书，版权所有，侵权必究。
京科版图书，印装差错，负责退换。